OFF TO SPACE!

PLANETS

by Rachel Rose

Consultant: Beth Gambro
Reading Specialist, Yorkville, Illinois

BEARPORT PUBLISHING

Minneapolis, Minnesota

Teaching Tips

Before Reading

- Look at the cover of the book. Discuss the picture and the title.
- Ask readers to brainstorm a list of what they already know about planets. What can they expect to see in this book?
- Go on a picture walk, looking through the pictures to discuss vocabulary and make predictions about the text.

During Reading

- Read for purpose. Encourage readers to think about the planets and what role they play in space as they are reading.
- If readers encounter an unknown word, ask them to look at the sounds in the word. Then, ask them to look at the rest of the page. Are there any clues to help them understand?

After Reading

- Encourage readers to pick a buddy and reread the book together.
- Find a place in the book that tells about planets and the sun.
- Find a page that tells what a planet is made of. Find a page that tells about a planet made of something else.
- Ask readers to write a sentence and draw a picture of something that they learned about planets.

Credits:
Cover and title page, © Vadim Sadovski/Shutterstock; 5, © Juan Lopez/Dreamstime; 7, © rwarnick/iStock; 8, © FlashMyPixel/iStock; 9, © NASA/JPL; 10, © Zakharchuk/Shutterstock; 12, © Dotted Yeti/Shutterstock; 13, © Wikimedia Commons/NASA; 15, © Tristan3D/Shutterstock; 16, © MarcelClemens/Shutterstock; 18, © Forplayday/Dreamstime; 21, © Zakharchuk/Shutterstock; 23BL, © LYagovy/iStock; 23BM, © suman bhaumik/iStock; 23BR, © Science Photo Library/Alamy Stock Photo; 23TL, © NASA/JPL; 23TR, © Volodymyr Goinyk/Shutterstock

Library of Congress Cataloging-in-Publication Data

Names: Rose, Rachel, 1968– author.
Title: Planets / Rachel Rose ; consultant, Beth Gambro, Reading Specialist.
Description: Minneapolis, Minnesota : Bearport Publishing Company, [2021] |
Series: Off to space! | Includes bibliographical references and index.
Identifiers: LCCN 2020042142 (print) | LCCN 2020042143 (ebook) | ISBN
9781647475079 (library binding) | ISBN 9781647475123 (paperback) | ISBN
9781647475178 (ebook)
Subjects: LCSH: Planets—Juvenile literature.
Classification: LCC QB602 .R654 2021 (print) | LCC QB602 (ebook) | DDC
523.4—dc23
LC record available at https://lccn.loc.gov/2020042142
LC ebook record available at https://lccn.loc.gov/2020042143

For more information, write to Bearport Publishing, 5357 Penn Avenue South, Minneapolis, MN 55419.

Printed in the United States of America.

Contents

We Live on a Planet!

Our world is a big place.
It is a **planet** called Earth.
Earth spins in space.

Earth moves around the sun.

But it is not the only planet to do this.

Seven other planets do, too!

The path they take is called an **orbit**.

Earth

Mercury is the planet closest to the sun.

It is also the smallest planet.

Next is Venus.

Venus is the hottest planet.

Mercury and Venus are made of rock.

So is Earth.

Earth is the only planet known to have life on it.

Mars has red rocks and dust.

It is called the Red Planet.

Mars has a huge **volcano**, too!

Then comes Jupiter.

Jupiter is made of gas.

It is the biggest planet.

More than 1,000 Earths could fit inside Jupiter!

Jupiter
Earth

Rings

Saturn is also made of gas.

Many **rings** circle around the planet.

The rings are made of rocks and ice.

Uranus and Neptune have lots of ice.

They are far away from the sun's heat.

Neptune is the last planet from the sun.

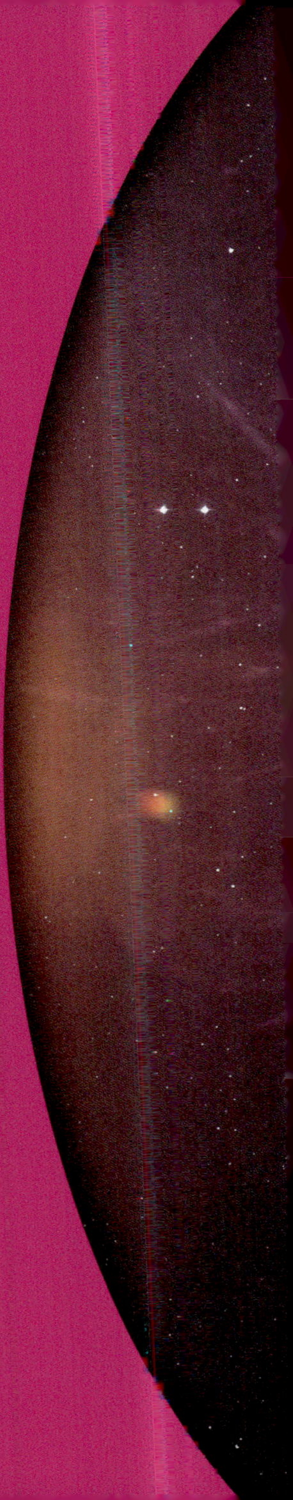

The sun and planets are part of our **solar system**.

But there are many other planets in space.

Maybe one day we will learn about them, too.

Fun Planet Facts

The planets in our solar system are over 4.5 billion years old.

Earth takes 365 days to orbit the sun. That is one year. It takes Neptune 165 Earth years to orbit the sun!

Some planets have moons. Earth has one moon. Saturn has 82!

Glossary

orbit the path of an object that is circling the sun

planet a large, round object that circles the sun

rings flat discs that are around some planets

solar system the sun and everything that circles around it, such as planets

volcano a mountain that can erupt

Index

Read More

Bell, Samantha. *Solar System (My World of Science).* Ann Arbor, MI: Cherry Lake Publishing, 2018.

Gillespie, Katie. *Planets (Eyediscover).* New York: Weigl, 2018.

Learn More Online

1. Go to **www.factsurfer.com**
2. Enter "**Space Planets**" into the search box.
3. Click on the cover of this book to see a list of websites.

About the Author

Rachel Rose lives in California. She thinks there's no better place to be than planet Earth.